A DAY IN SPACE

by Suzanne Lord and Jolie Epstein

SCHOLASTIC INC.
New York Toronto London Auckland Sydney

Photographs by NASA
Art direction by Diana Hrisinko
Cover art by Walter Wright
Photo design by Sue Ewell
Text art by Erich Barnes and Hima Pamoedjo

ISBN 0-590-40152-1

12 11 10 9 8 7 6 5 4 3 2 1 4 6 7 8 9/8 0 1/9

Printed in the U.S.A. 24

What would it be like to go up in a NASA Space Shuttle? How would you feel? What would you do in a day?

Astronaut Jeff Hoffman knows because he's been there. He'll tell us how he felt.

But what about you? There's only one way to find out. So let's go!

First you must go through NASA's astronaut training program.

"There is a lot to learn," Jeff Hoffman says. "They pile books from ceiling to floor! There's a lot of studying!"

You learn about science, space, and flying. And of course, you learn how to use your equipment!

What better way to learn about floating in space than by floating in water! That's how you learn to move around in your space suit.

You learn to use the shuttle's control board by working on an exact model of it. This model is called a simulator.

It's the real thing—lift-off time! Are you ready? You look it. But even with all your training, you feel a little odd.

"You have to be a little bit nervous about sitting on top of a rocket that's getting ready to launch," says Jeff.

Does it help to know that everyone on the crew feels the same way?

The countdown begins, and suddenly there is a lot of noise.

"There's a big bump when you first light up the engines, so you can tell you're on your way," Jeff says. "Then everything is bright yellow from the reflection of the fire! Then you're shaking up and down and back and forth! But that only lasts a couple of minutes."

As the rocket goes up, you are pushed down into your seat. This is called a G-force.

"You can feel yourself being pressed down. You actually weigh three times as much during a launch as you do on earth," Jeff says.

Suddenly, you are not pressed down anymore. You feel as light as a feather.

Surprise! You are as light as a feather. In fact, you are weightless!

"When that happened," says Jeff, "I knew I was in space!"

Remember that pencil you put down just before the launch? It just floated past your head.

And that's not all. Any dust you brought in will float, too. Don't cry! Your tears will float around. And please try not to sneeze!

Luckily, there is a filter system that keeps the air as clean as possible.

The living and working quarters on the shuttle have air. So you don't have to wear a space suit there. But right now, your clothing is about the only normal part of you!

"Weightlessness is a little uncomfortable at first," says Jeff. "Your stomach feels the way it does on an elevator. And your head feels the way it does when you hang upside down!"

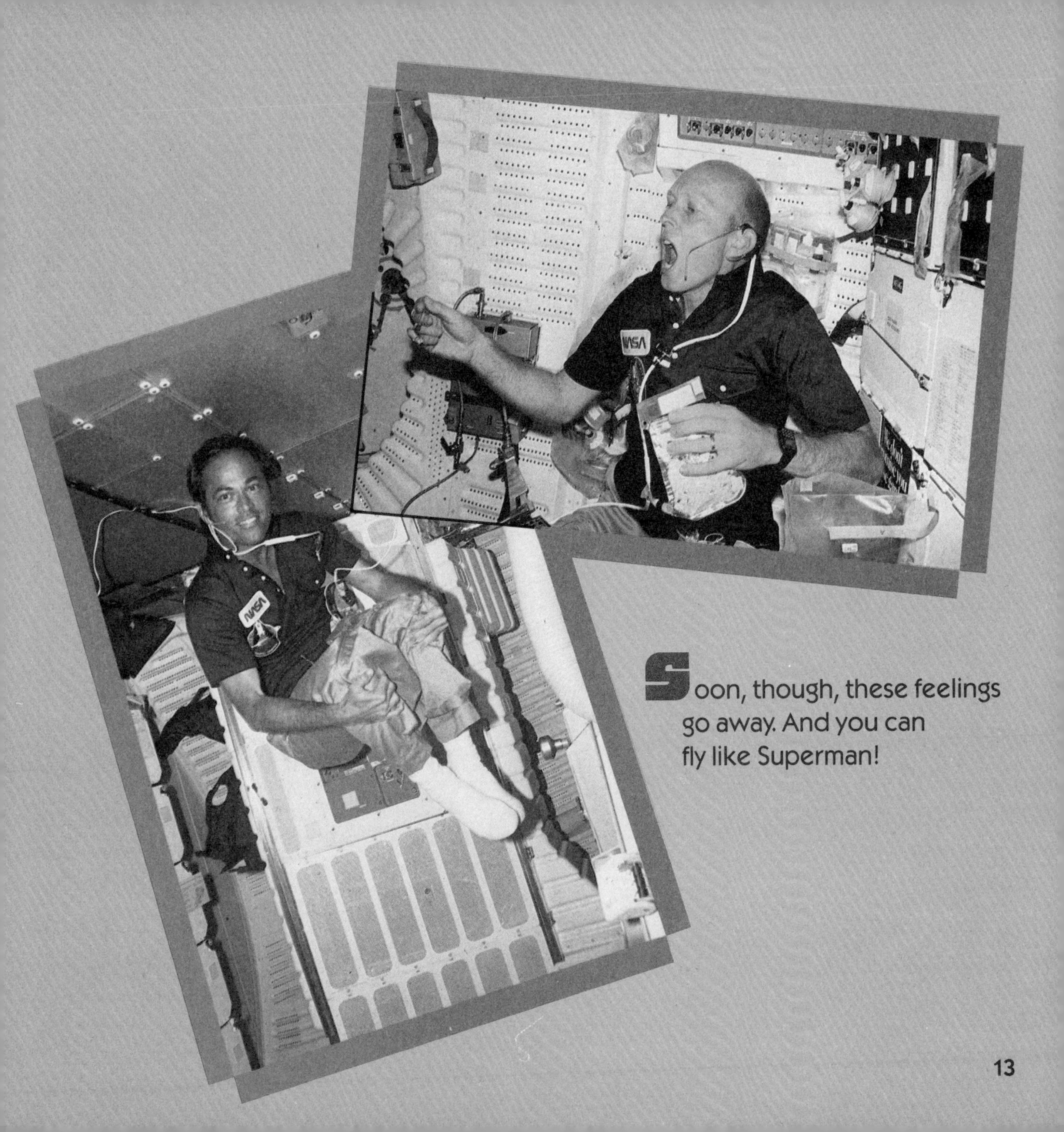

Soon, though, these feelings go away. And you can fly like Superman!

What's for supper? Maybe some smoked turkey, mixed vegetables, strawberries, and cream of mushroom soup. Happy appetite!

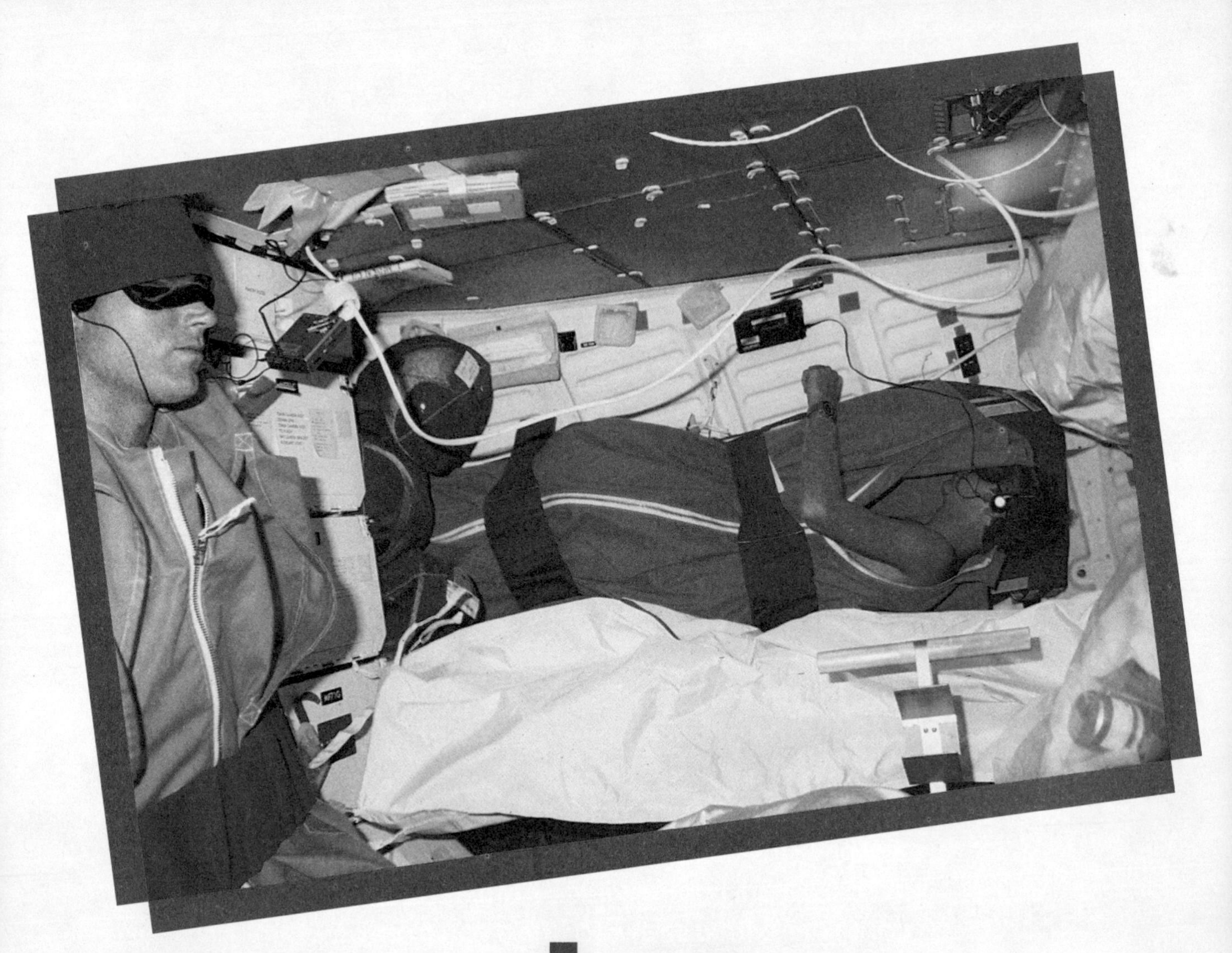

It's already time for bed. As you drift off to sleep, you listen to music through headphones. Sleep tight.

Good morning! Are you ready for your first workday? You're getting used to floating around, and right now you float to breakfast!

Food is a little different here. Some of it—like peanuts, Life Savers, and fresh fruits—is normal. But most of your food is dehydrated. It's all dried up!

"It's camping food," says Jeff. "You add water and heat it. But you have to make sure that it's very sticky, so that it sticks to the bowls. Otherwise, it would go floating all over!"

You have Instant Breakfast, a piece of fresh fruit, and orange juice from a special container with a straw. To keep you from floating all over, your feet are in footholds!

After breakfast, you wash up. Toothpaste squeezes onto the toothbrush normally. But you can't just spit! You must swallow the toothpaste or spit it into a towel.

There is no shower. The water would float all over the place in wet globs! So you bathe with a washcloth. There is a special place for you to put your hands if you need to wash them. You have to strap yourself to the toilet, where a suction system is used to "flush" wastes!

The crew has just received its daily work schedule. Every morning, NASA's ground crew tells the shuttle members what they are supposed to do for that day.

The shuttle is carrying a new satellite. Your job is to take it out of the cargo bay and send it into orbit around the earth.

FLIGHT DIRECTOR

The cargo bay is open already. That was done just after the launch to cool the ship down.

The satellite is held in with protective covers. These covers and a robot arm are controlled from the desk.

From the desk, you open the covers. The satellite starts to float out. You get the robot arm to move the satellite away from the shuttle.

This satellite has its own rockets. NASA ground control will set these rockets off. Then the satellite will fly far away into its own orbit.

Wait—something is wrong! The satellite's rockets didn't fire!

NASA ground control says a switch on the side of the satellite didn't go on. That's why the rockets didn't fire. Can the shuttle crew get the switch to flip on somehow?

The crew is worried. It won't be easy.

You decide to try to flip the satellite's switch on with the robot arm. First, you rig up an attachment for the arm. You call it a fly swatter.

It looks a little like a fly swatter. You hope the switch will get caught in one of the swatter holes. Then the robot arm can move the switch.

When the swatter is finished, two crew members must attach it to the end of the robot arm. The robot arm is outside.

You are going on a space walk!

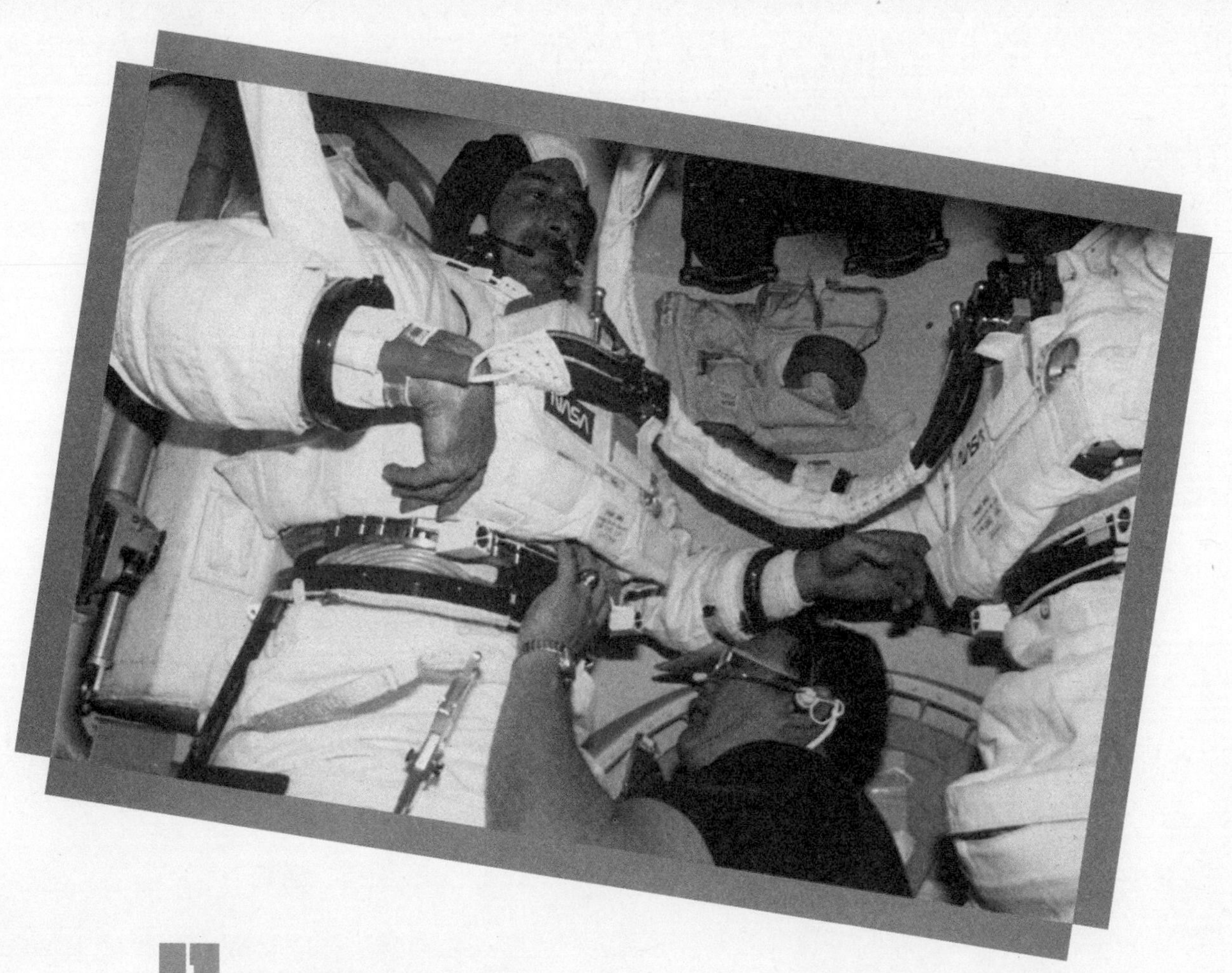

Your space suit comes in three parts: pants, top, and helmet. Nowadays it only takes about five minutes to get into it. In the early days of space travel, it used to take an hour!

NASA

Are you ready? You know there is air in your helmet. But even so, you hold your breath as you step out.

Look down. What do you see under your feet? That's right—

NOTHING!

Guess what you are right now? A satellite! You are in orbit, too.

You come back from your space walk and the robot arm is ready for its next task.

Your shuttle and the satellite are orbiting at different speeds. You have to wait until they are close again.

Here it comes, turning over and over again. You send the robot arm out. It misses the switch! The satellite turns over again. You try again. Another miss!

But the third time, your fly swatter grabs the switch. You move the arm. The switch flips.

Nothing happens! The rockets still don't send the satellite into its own orbit far away!

Ground control says that something else must be wrong. What do you do next? NASA says forget it. Maybe the satellite can be salvaged on another trip.

Oh, well. Not every mission can go perfectly!

Your next job might be to check on the experimental animals on board the shuttle. "Most of the animals are rats," Jeff explains. But not all of them are!

"We took flies," Jeff says. "It took them a couple of days to learn to fly around in space!"

Not all experiments include animals.

"There are a lot of things you can make without gravity," says Jeff.

Oil and water don't mix on earth—but they will in space, because they are both weightless. The same thing happens with a lot of things.

"You can make types of medicines up here that you couldn't make on earth," Jeff says. "Probably in ten years, people will be taking special medicines that were made in space!"

Not all experiments are serious, either. Part of your work might be to see how toys work in space!

"Yo-yos work great," says Jeff. "They go very slowly, so you can actually watch what it's doing and control it.

"And jacks! You bounce the ball off the ceiling and try to get the jacks before they float away!

"But marbles—they 'catch' each other and form little circles. Beautiful!"

What do you do in your spare time? Read? Play card games with the other crew members?

Not really.

"Everyone looks out the window," says Jeff. "I can read anytime. But where else can I look at the world from the height of two hundred miles?"